THE THEORY OF ONE

REALIZING THE DREAM
OF A FINAL THEORY

Christopher Bek

 FriesenPress

Suite 300 - 990 Fort St
Victoria, BC, Canada, V8V 3K2
www.friesenpress.com

Copyright © 2015 by Christopher Bek
First Edition — 2015

ISBN
978-1-4602-7554-2 (Paperback)
978-1-4602-7555-9 (eBook)

1. Science, Physics

Distributed to the trade by The Ingram Book Company

www.philosophymagazine.com

TABLE OF CONTENTS

INTRODUCTION

The speed of light in a vacuum is one of nature's few universal constants. Special relativity theory established that the speed of light is the universal speed limit. No material object can actually reach this speed. Since any object gains apparent mass as it goes faster, gaining an infinite amount at the speed of light, it would take an infinite amount of energy to accelerate to this speed. By the same token, this would become zero if light could slow down.

—*Ian Marshall and Danah Zohar*

Planck's constant is one of the two most important constants in the whole of modern physics, the other being the speed of light. Max Planck was one of the early founding fathers of quantum physics. His main contributions were the theory that electromagnetic radiation happens in discrete quanta, and the discovery that the size of each quanta is associated with a universal constant, a physical ratio or proportion that stays the same in all circumstances and in all frames of reference.

—*Ian Marshall and Danah Zohar*

I want to know God's thoughts. The rest are details.

—*Albert Einstein*

REPETITION OF IDEAS

It has been pointed out to me that I tend to repeat myself.
I do this for four reasons. Firstly, it is good to see ideas
(like Schrödinger's cat and Pascal's sphere) in different contexts. Secondly, some readers need to see ideas
presented more than once. Thirdly, once I have defined
pi it cannot be redefined. I take my ideas to the apex of
development the first time and thus cannot do any better.
Fourthly, very few books even present simple and beautiful ideas at all. I would argue that it is better to see them
several times than not at all.

THE CARTESIAN METHOD OF READING

René Descartes (1596–1650) was a French scientist,
mathematician and the father of modern philosophy. He
began his philosophic quest for certainty by tearing down
the medieval house of knowledge and then rebuilding it
from the ground up. Descartes employed the method of
radical doubt when asking the simple question, "What
do I know for certain?", to which he concluded that
he certainly knew of his own existence, an insight he
immortalized with his celebrated *Cogito, ergo sum,* "I
think, therefore I exist." Following Descartes' lead, *The
Theory of One* takes the reader on a fantastic quest for
cosmic certainty. Here, the responsibility of the reader
lies in simply embracing the theory as the null hypothesis
for the brief time it takes to read this essay. After accept-
ing the theory for a time, the prudent reader will be
positioned properly to accept or reject the theory with

certainty once and for all. Descartes expressly advocated the option-based approach to reading. Accordingly, the first pass is strictly experiential—as if driving a convertible sports car along a mountain road on one of the last days of summer. The goal here is simply to keep the car on the road. The second pass requires a more careful reading and asks that the reader mark pertinent passages and make plenty of notes in the margins. The third and final pass calls for the rereading of notes and marked passages. The Cartesian method provides the option of forgoing the second and third readings while, at the same time, affording the reader a good basic sense of the book. This book was written carefully for the cultured public. It is a ninety-minute luxury vacation through the brave new world of "onespace" for those who have taken the road less traveled. For the sake of convenience, the book has been designed specifically to fit comfortably into an inside coat pocket.

Rebooting the Machine

SUMMARY → "Rebooting the Machine" argues that we must reboot all societal machines and then rebuild from the ground up. We could start this rebooting with the discipline of physics. As such, I am asking the director of the Perimeter Institute of Theoretical Physics to respond to my theory of one.

Canada's premier science institute is trying to jumpstart a revolution in physics, in part by encouraging the randomness of human brilliance. For inspiration on the way forward, the Perimeter Institute is looking 100 years into the past.

—*Ivan Semeniuk*

O N MY BIRTHDAY in 2000, I smoked a joint and read Edgar Allan Poe's 1848 classic, *Eureka*. In it, Poe refers to Blaise Pascal's claim that the universe is a sphere in which the centre is everywhere and the boundary is nowhere. Two days earlier I had read Rudy Rucker's 1977 classic, *Geometry, Relativity and the Fourth Dimension*, in which he showed Einstein's proof that gravity and inertia are mathematically equivalent. I took Pascal's sphere and inverted it so that the universe is a sphere in which the boundary is everywhere and the centre is nowhere. Following Einstein, I then proved that Pascal's sphere and the inverted Pascal's sphere are mathematically equivalent. After that I replaced the words boundary and centre with the words lightspeed and Planck's constant—thereby uniting relativity theory (based on lightspeed) and quantum theory (based on Planck's constant). As Einstein said, "This is so simple God could not have passed it up." That night I met a friend and told her about my amazing discovery. This cosmic truth owes its existence in part to marijuana, which enables me to reboot my mind and reconceive the universe.

REBOOTING THE MACHINE

In its simplest terms, rebooting tells us to turn the machine off and then on again. It is a healthy refresher. René Descartes (1596-1650) rebooted knowledge by tearing down the Medieval house of knowledge and rebuilding from the ground up. Government elections are a form of rebooting. When we clean out our fridge we are performing a type of rebooting. Rebooting is part of the natural cycle of life. It is necessary and sufficient to keep our civilization moving forward. Rebooting jump-starts both evolution and revolution.

STAIRWAY TO HEAVEN

Ontology is the branch of philosophy that addresses the question of what are the fundamentally distinct levels of being that compose reality. The four dimensions of ontology are matter, life, consciousness and self-awareness. God and souls are self-aware. According to EF Schumacher,

> From a base of matter, man has the power of life like plants, the power of consciousness like animals and the power of consciousness recoiling upon itself. This power of self-awareness opens up unlimited possibilities for the purposeful learning, formulating and accumulating of knowledge. Scientists tell us that we cannot talk about a life-force because no such force has ever been found to exist. Yet the difference between alive and dead certainly exists; and the same goes for consciousness and self-awareness.

Rebooting simply requires that we go back to basics and climb the stairway to heaven so that Being can erupt. And similar to regular ontology, the ontology of science breaks down into a base of mathematics, then physics, chemistry and biology.

THE PhD DISEASE

Just as Galileo and Descartes laid down the modern scientific method for solving problems—so too did Harry Markowitz in 1952 lay the modern scientific method of bringing together an uncertain set of variables. It is focused on portfolio distributions and risk-reward efficiency. The portfolio distribution describes the uncertainty of future outcomes. Risk-reward efficiency means that with hi/low risk comes hi/low reward. But rather than seeing portfolio theory for the important roadmap that it is—it has been pedantically attacked by PhDs like the legions of PhDs at Enron before the company finally collapsed under the weight of its own PhDness. Similarly, theoretical physics PhDs have built their castles on sand. They should consider rebooting theoretical physics by conducting phenomenological examinations of space-time, atoms and light as I did with my theory of one. A phenomenological examination of a glass of beer might go something like this—An amber-colored liquid with bubbles that is generally considered to taste good and has an intoxicating effect.

REBOOTING PHILOSOPHY

Western philosophy effectively began with the Greek Plato (427-347 BC) who dared to ask what existence would be like outside the cave. Platonic thought serves to inform and support those who have taken the road less travelled in the struggle towards daylight. Alfred Whitehead argued that all philosophy after Plato is merely a footnote. René Descartes (1596-1650) was a French scientist, mathematician and father to modern philosophy. He began his philosophic quest for certainty by rebooting the medieval house of knowledge and then building from the ground up. Descartes employed the method of radical doubt when asking the simple question—What do I know for certain?—to which he concluded that he certainly knew of his own existence—which he then immortalized with his celebrated *cogito*—ie. *Cogito, ergo sum*—I think, therefore I exist. With this, Descartes rebooted modern philosophy and science while I am now rebooting postmodern philosophy and science based on transitions like from behaviorism to existentialism, and like from string theory to my theory of one. Following Plato, I take the best of all philosophers and scientists and bring them together into a holistic, unified body of knowledge.

REBOOTING PSYCHIATRY

Existentialism is the philosophy that stresses individual existence, freedom and choice. It views humans as defining their own meaning in life as beings who try to make rational decisions in an irrational world. Kierkegaard, Nietzsche, Dostoyevsky, Camus, Kafka and Sartre are among the many great existential philosophers. Existentialism is all about putting man in touch with himself. It stresses that individuals have total freedom and total responsibility for the world. Alternatively, behaviorism is the psychological model employed today. It tells us we have no freedom and no responsibility other than to behave ourselves. The absence of freedom stems from the worldview of determinism, which says the question of whether we choose to walk the dog tonight was determined at the moment of the Big Bang (i.e. the creation of the universe) sixteen billion years ago. We could reboot psychiatry by switching from behaviorism to existentialism. All that is required is for psychiatrists and educators to teach existentialism to their patients and students.

REBOOTING GOD

We could say that the Big Bang is occurring at every moment going back to its origin sixteen billion years ago when a photon (i.e. a being of light) splits into an electron and a positron (i.e. matter and antimatter). By definition, photons travel at lightspeed and thus exist at the boundary of the universe. This universal boundedness leads to the realizations that there is only one photon and that photon is God. From outside the universe, a single photon appears as a spherical film containing the universe—like a translucent pearl encapsulating a grain of sand. One might even argue that God and light are the same thing. It also follows that the universe exists inside of God's womb (ie. Her womb). In essence, God and the universe therein contained can be effectively seen as a single particle. Sir James Jeans (1877-1946) said that God is a mathematician. Niels Bohr (1885-1962) defined the complementary principle as the coexistence of two necessary and seemingly incompatible perspectives of the same phenomenon. Therefore, God is both the photon and a mathematician. Saint Augustine (354-430) portrayed existence as an ontological set of stairs leading to God. Augustine also said God exists outside of time and the universe was created with time and not in time.

THE PERIMETER INSTITUTE

According to the Perimeter Institute of Theoretical Physics,

> We are an international focal point of cutting-edge research in foundational theoretical physics. The Perimeter Institute is a leading centre for scientific research, training and educational outreach in theoretical physics. Founded in 1999 in Waterloo, Canada, our mission is to advance our understanding of the universe at the most fundamental level. Perimeter also trains the next generation of physicists through innovative programs, and shares the excitement and wonder of science with students, teachers and the general public. We believe breakthroughs are realized through a collision of intellect, imagination and inspiration. The Perimeter Institute is an independent, resident-based research institute devoted to foundational issues in

theoretical physics at the highest levels of international excellence. The Perimeter's vision is to create the world's foremost centre for foundational theoretical physics, uniting public and private partners, and the world's best scientific minds, in a shared enterprise to achieve breakthroughs that will transform our future.

REBOOTING PHYSICS

In a recent article by Ivan Semeniuk of The Globe and Mail newspaper entitled, *A Calculated Reboot*, he spoke with the director, Neil Turok, of the Perimeter Institute. The article states that the institute is trying to perform a calculated reboot of theoretical physics and turn the clock back a hundred years to relativity theory (1905) and quantum theory (1925). I am now announcing to the world of physics that I have rebooted theoretical physics and have turned the clock back one hundred years with my theory of one (2001). With it, I have delivered the ocean to Turok. In formulating my theory, I ignored all theoretical physics after the Schrödinger's cat thought problem (1935). I rebooted physics by focusing on ontology and have reexamined relativity theory and quantum theory at the most fundamental level. Relativity theory is the natural law of spacetime and is based on lightspeed. Quantum theory is the natural law of the atom and is based on Planck's constant. The theory of one unites the two theories by proving that lightspeed and

Planck's constant are the same boundary of the spacetime continuum (ie. the universe).

Conclusion

We must be prepared to reboot absolutely everything in order to achieve breakthroughs that will transform our future. And as my contribution to human brilliance, I first produced my theory of one on 1 January 2001 based on one argument—and have since added six more arguments. I now invite the director of the Perimeter Institute to argue either for or against my theory. He is encouraged by the laws of nature to write an essay that responds to my theory of one. As Edmund Burke (1729-97) said, "The only thing needed for evil to triumph is for good men to do nothing."

AGAINST PHYSICS

SUMMARY → "Against Physics" recounts the two major physical theories developed during the 20th century in the context of Ockham's principle of economy and Dirac's principle of aesthetic value.

It is more important to have beautiful theories and equations than to have them fit the data.

—*Paul Dirac*

ONCE, I PRESENTED a lecture on time-series risk modeling at Cornell University in Ithaca, New York. While traditional risk metrics represent risk over a single time period (e.g. one day), time-series risk modeling represents uncertainty over multiple time periods (e.g. thirty times one day). Time-series risk modeling enables us to look much further into the future. The method capitalizes on the conceptual similarities between space and time and is, in essence, a temporal telescope. My approach breaks down historical data (e.g. daily oil prices) into signal, wave and noise. I forecast the three components separately and then recombine them into a single probability distribution like the normal distribution using Monte Carlo simulation. The approach effectively rolls the dice over and over in order to bring all the moving parts into a single distribution. Monte Carlo simulation is an advanced method that it can handle any number of input variables and combine them into a single output distribution, which can then be used to make efficient risk-reward decisions going forward. Time-series risk modeling turns uncertainty into opportunity and is undeniably destined to become the favorite toy in the CFO's toy box.

TIME TO A PIG

I awoke early one Sunday morning and caught a plane from Calgary to Toronto. From there, I rented a car and drove through Buffalo and then east along the interstate and finally south alongside Lake Cayuga towards Ithaca. While exploring some of the spectacular wine country around the lake, I came across a farmer standing under an apple tree with some pigs. I watched the farmer while he held a pig above his head and the pig ate an apple from the tree. He put the pig down and picked up another pig, only to perform the same mesmerizing act once again. I approached the farmer and asked him whether he might be better off taking some of the apples from the tree and putting them on the ground for the pigs to eat there, which might save time—for which he responded, "What's time to a pig?" The point of the story is that the farmer could only see reality through the eyes of the pig—just like most people can only see reality through the eyes of the government.

OCKHAM'S RAZOR

The English monk William of Ockham (1285–1349) was one of the most important thinkers of all time. He was noted for his keen sense of logic and his enduring theo-logical ideologies. Going entirely against the traditional philosophical reasoning of his day, Ockham put forth his now famous principle of economy, which states that a plurality of reasons should not be postulated without necessity. In other words, all things being equal, the simplest theory tends to be the correct one. Ockham used his principle so frequently and with such purpose that it became known as "Ockham's razor." He claimed it was vain to do with more what could be done with less. Even today, Ockham's razor remains the foundation of all truly authentic scientific reasoning.

AGAINST COMPLEXITY

In keeping with the principle of economy, the young physicist Albert Einstein (1879–1955) was said to have so totally despised complexity that he refused to buy shaving soap when regular washing soap was sufficient. At the time, Einstein and others were working toward resolving the conflict between the newly realized invariance of light speed and absolute Newtonian space. Einstein unraveled the conundrum in 1905 with his special relativity theory by combining the separate notions of relative space and relative time into absolute spacetime. Einstein invoked Ockham's razor again in 1915 when postulating his general relativity theory, based on the revelation that gravity and inertia are the same thing.

ROTATING THE AXES

In characterizing relativity theory, consider for a moment an astronaut named Elvis traveling through space. Consider also a two-dimensional representation of the spacetime continuum with time indicated on the vertical axis and distance indicated on the horizontal axis. According to this application of the Pythagorean Form, as Elvis accelerates toward light speed (i.e. 186,284 miles per second), his spacetime axis rotates clockwise so that the time axis moves toward the original distance axis. If Elvis were traveling at half of light speed, he would be only 87 percent of his original height, while time for Elvis would elapse at 87 percent of its original rate. If Elvis were to achieve light speed somehow, his spacetime axis would rotate a full ninety degrees from its original position, and he would disappear from the spacetime continuum altogether. In other words, upon reaching light speed, we could say that, "Elvis has left the universe."

What's Spacetime to a Photon?

It would, of course, be impossible for Elvis to actually achieve light speed, given that such an endeavor would require all of his mass be converted into energy in accordance with Einstein's famous $E = mc^2$ equation. However, it is interesting to note that, by definition, beings of light or *photons* travel at light speed—thus implying that photons must exist at the boundary of the spacetime continuum (i.e. the universe). As such, one can imagine the universe populated by a fantastic number of photons zipping around the periphery. And if one were to choose this moment to invoke Ockham's razor, the obvious question would be, economically speaking, why would the universe need more than one photon if photons exist outside of spacetime? The answer to such a question would certainly be that there does not need to be more than one photon—from which we must therefore conclude that there is but one photon, which encapsulates the physical universe. Without light (i.e. the lone photon) we would all be looking for our keys in the dark.

Authenticity

The British physicist Paul Dirac (1902–1984) was just twenty-six years old when he formulated the relativistic wave equation that led to the discovery of antimatter in the form of the antielectron or *positron*. Dirac discovered the key when he realized that $E = -mc^2$ is a valid equation as well. Interestingly, Einstein was also twenty-six when he put forth special relativity and, like Dirac, worked alone for his entire career. The aesthetically-deduced relativistic wave equation became a cornerstone of quantum theory in that it integrated Einstein's relativity theory into Schrödinger's wave equation, thus accounting for the behavior of electrons traveling near light speed. In putting his own enlightened spin on Ockham's razor, Dirac claimed it is more important to have beautiful theories and equations than to have them fit the data. While lesser philosophical disciplines like empiricism and positivism have attempted to deploy Ockham's razor as the basis for systems aimed at producing totally objective reasoning, Dirac found the missing link in subjectivity, which enabled scientific reasoning to make the next quantum leap forward. To be sure, genuine art is always economical, but its true value can only be arrived at subjectively. Interestingly, the American physicist John Wheeler once subjectively asserted that positrons are nothing more than electrons traveling backwards in time.

Through the Looking Glass

Dirac shared the Nobel Prize with Erwin Schrödinger (1887–1961), best known for studying the wave mechanics of orbiting electrons and helping to formulate quantum theory. In attempting to explain the paradoxical nature of quantum theory, Schrödinger put forth the classic "cat in a box" thought problem as follows. A "quantum cat" (let's also call him Elvis) is placed in a box. The box is such that no one can know what is happening inside. A device triggers the release of either food or poison with equal probability, and Elvis meets his fate—or does he? In the strange world of quantum theory, subatomic particles exist in several places at once and only become determinate upon observation. In other words, consciousness determines physical reality. Therefore, Schrödinger argued that Elvis is both alive and dead until the moment the box is opened. Inside the box, unobserved, his state of existence can only be described by a probability distribution. Remarkably, quantum theory claims universal indeterminacy beyond the boundary of Planck's constant. In fact, Planck's constant is a threshold of the universe in the same way that light speed is a threshold. While in the unobserved state, Elvis rotates out of the spacetime continuum into a metaphysical realm where determinism is replaced by indeterminism—and where we can say once again that "Elvis has left the universe." Upon opening the box, the observer's consciousness becomes present, and Elvis snaps back into the spacetime continuum, whereupon his fate is summarily determined. Determinism is the worldview that the decision as to whether or not to floss your teeth tonight was made at the moment of the Big Bang sixteen billion years ago.

ROTATING OUR MINDS

In his 1992 book, *Dreams of a Final Theory,* Nobel Prize-winning physicist Steven Weinberg includes a chapter entitled "Against Philosophy," in which he claims that, historically speaking, philosophy has contributed little to our understanding of physical phenomena. According to Weinberg,

> Most physicists carry around a working philosophy, a rough-and-ready realism, a belief in the objective reality of the ingredients of our scientific theories. But this has been learned through the experience of scientific research, and rarely from the teachings of philosophers.

CONCLUSION

One is presently reminded of the ancient story of the three monks who happened across a flag blowing in the wind. The first monk claimed the flag was moving. The second monk claimed the wind was moving. The third monk then trumped the first two by claiming that it was only our minds that were moving. And strictly as a matter of interest, one wonders what the third monk would have to say about the lone photon moving across the sky at light speed. The simplest and most aesthetically appealing theory would be that God and light are the same thing. As the deeply religious Einstein remarked, "I fully expect to spend the remainder of my life pondering the nature of light."

TRANSCENDING UNCERTAINTY

SUMMARY → "Transcending Uncertainty" recounts the events leading up to the paradigm shift of quantum theory in 1925 and then takes a look at what we still have to learn from it. The nanosecond forecast of Philosophymagazine.com calls for a monumental paradigm shift, whereby we will finally orient ourselves to the universe.

We dance around in a ring and suppose
while the secret sits in the middle
and knows.

   —Robert Frost

A FRANK & ERNEST comic strip shows a chick breaking out of its shell and then looking around and proclaiming, "Wow, paradigm shift."

NO PAIN, NO PAIN

Thomas Kuhn (1922–1996) was a physicist and historian concerned with the sociology of scientific change. In his 1962 book, *The Structure of Scientific Revolutions,* he defines the term *paradigm shift* as a transformation that takes place beyond the grasp of our undeveloped *a priori* comprehension. *A priori* is Latin for "before experiment" while *a posterori* stands for "after experiment." Kuhn also defined normal science as *a posterori*, meaning well established science, like Newtonian physics, which is the basis of our civilization. The point is that such *a priori* or nonlocal shifts cannot be achieved via stepwise, linear thinking. Typically, scientists apply normal scientific methods within a paradigm until the paradigm weakens and a shift occurs. Most physicists eat up normal science with a big spoon but do everything possible to avoid the intense metaphysical pain of paradigm shifts. Philosophy serves to temper the pain by providing higher vantage points so as to better prepare us for impending shifts. Notably, the markets, evolution, consciousness, and virtually every single phenomenon in the universe exhibits this same fractal pattern—as characterized by long periods of stagnation followed by dramatic, nonlocal jumps. Fractal means "fractions of dimensions" and contends that there is no inherent scale to the universe. So, metaphorically speaking, there is no difference between looking through a telescope and looking through a microscope. The telescope represents relativity theory while the microscope represents quantum theory.

FAIT ACCOMPLI

All indications suggested that the world of physics was on the verge of completion 120 years ago. Comprehensive theories were in place to describe the two known universal forces—gravity and electromagnetism. Sir Isaac Newton's (1642–1727) laws of gravitation and motion effectively characterized the mechanics of all physical bodies, and the differential equations of James Maxwell (1831–1879) portrayed the wave mechanics of visible light and other electromagnetic forces. At the time, scientists believed the universe was a deterministic, clock-like machine that followed strict causality. In support of this belief, Pierre Laplace (1749–1827) claimed once that, deterministically speaking, we should be able to predict absolutely everything in the future or the past given detailed knowledge of the present.

SPLITTING THE ATOM

Classical physics and determinism started to unravel in 1897 when Sir J.J. Thomson (1856–1940) discovered the electron and the divisibility of the seemingly solid atom. He advanced the "plum pudding" atomic model by suggesting that electrons were like plums in a pudding of positive matter. Three years later, Max Planck

(1858–1947) made the groundbreaking discovery that energy is transferred in discrete packets or *quanta*, as defined by Planck's constant. For example, if we had a Planck's constant of beer, we could only drink all or none of the beer and nothing in between. The revelation of this new paradigm signifying nature's ultimate granularity seemed so outrageous to Planck that he spent years looking for a more conventional rationale.

WAVE DISPLACEMENT

Einstein produced three papers in 1905. *The Photoelectric Effect*, for which he was awarded the Nobel Prize, applies the quantum concept in suggesting that light is also transferred in discrete packets known as *photons*. This *quantizing* of light presented a problem in that electromagnetic mechanics unequivocally describes light as being wave-like. In fact, Einstein was the first to accept the seeming contradiction that light is both wave-like and particle-like. Einstein's second paper, *Brownian Motion*, delineates the stochastic process (i.e. skillful aiming in predicting future events) and is the very foundation of all risk modeling today, which is the mathematical representation of uncertainty that sometimes employs Monte Carlo simulation. His third paper, *Special Relativity*, reveals the interrelation of space and time, the interrelation of energy and matter, and the linear unification of Newtonian and electromagnetic mechanics. Notably, Einstein's approach was always to seek elemental conceptual pictures before considering mathematical complexities.

The Quantum Leap

Ernest Rutherford (1871–1937) proposed a solar system atomic model in 1911 based on the revelation that both the solar system and the atom have nuclei containing about 99.9 percent of the mass and occupying about one-billionth of the spherical space. Rutherford's assistant, Niels Bohr (1885–1962), realized that electrons are held in orbit electromagnetically rather than gravitationally. Bohr followed Einstein's lead in quantizing Rutherford's atomic model so as to produce a model with discrete electron orbits similar to what is taught in high school chemistry as *valence rings.* In conventional terms, valence rings represent the circular paths on which electrons orbit the nucleus of the atom. Prince Louis de Broglie (1892–1987) argued that matter also has both wave-like and particle-like properties. In 1925, Schrödinger con-structed an atomic model based on de Broglie's concept of matter waves while Werner Heisenberg (1901–1976) constructed a model based on matrices of infinite dimen-sion. Dirac then nailed down quantum theory once and for all by proving the two atomic models are equivalent. And, as the reader may know, televisions, computers and laser disk players are all based on quantum theory.

A Matter of Interpretation

While light waves are similar in nature to the waves that occur when dropping a pebble in the ocean, matter waves have no such physical meaning. It was Max Born (1882–1970) who proposed the idea that matter waves could be interpreted probabilistically. As such, the wave crests represent the highest probability of an electron landing on a given point on the wave—and coincides with the discrete electron valence paths of Bohr's model. Heisenberg captured the essence of this interpretation perfectly in 1927 with his famous "uncertainty principle," which states that causality breaks down at the boundary defined by Planck's constant. Quantum theory describes the microcosmos inside the atom and is the universal law of matter.

Causality

Causality is simply an ordering of spacetime. An absence of causality means an absence of spacetime, thereby indicating a boundary of spacetime. My theory of one argues that light speed and Planck's constant are the same boundary of the universe. It shows that there is spacetime inside the universe and nothingness both outside the universe and inside the atom. Schrödinger set forth his classic cat in a box thought problem in 1935 with the intention of demonstrating the absurdity of the probabilistic interpretation once and for all. The thought problem proves mathematically that the consciousness of the observer actually determines whether the cat is alive or dead. Eugene Wigner was one of the few physicists at the time concerned with the role that consciousness plays in determining physical reality. If consciousness determines physical reality, as the Schrödinger's thought problem implies, then the moon does not exist when no one is conscious of it. Even today, consciousness remains the secret in the middle.

EXPERIENCING VERTIGO

Special relativity (1905) shifted the cosmic paradigm from linear space to linear spacetime. General relativity (1915) shifted the paradigm from linear spacetime to curved spacetime. Quantum theory (1925) transcended material uncertainty and shifted the paradigm from determinism to indeterminism, which Einstein denied by claiming, "God does not play dice." Alas, it seems the capacity for paradigm shifts is limited in even the greatest of minds. As Einstein himself tells us,

> The years of searching in the dark for a truth that one feels but cannot express; the intense desire and alterations of misgivings; until the final break into daylight. This can only be known to those who have experienced it.

THE BOUNDARY VALUE PROBLEM

Saint Augustine (354–430) characterized existence as an ontological set of stairs leading to God. The four ontological steps are matter, life, consciousness, and self-awareness, which is the highest step reserved for God, souls and forms. Plato regarded forms as holding the highest level of being in the universe. Augustine maintains that God exists outside of time and that the universe was created with time and not in time. Later, Saint Thomas Aquinas (1225–1274) put forth an argument for the existence of God by asserting that every event is caused by a prior event that leads back to the first cause, which is God. But the argument falls down for the reason that causality is a temporal concept and, by definition, has no meaning outside spacetime.

NONLOCALITY

Nonlocal means "a jump of great distance." Paradigm shifts are nonlocal while normal science is local. Blaise Pascal (1623–1662) described the universe as a sphere in which the center is everywhere and the boundary is nowhere. Imagine the universe as a spherical ocean of spacetime with consciousness at the center like a newborn lying in a crib. Imagine photons surfing the waves of the ocean while electrons leap in and out like synchronized dolphins. Beyond the ocean is beyond our comprehension. Since both photons and dolphins exist at the boundary of our comprehension, it is reasonable to believe they appear to us sometimes as waves and sometimes as particles. Thus, quantum theory provides the best estimates as to the local and nonlocal reentry locations of the dolphins. And there is certainly no reason why the dolphins cannot reenter the ocean before their actual time of departure—that is, travel backwards in time. I would then argue that reality is presenting itself to our newborn minds by way of some very hardworking dolphins. And Pascal's sphere supplies the conceptual picture that

Einstein overlooked in claiming that God does not play dice. Consider also the conceptual picture of the universe as a pearl with the pearl substance as heaven housing God, souls and forms. The grain of sand in the middle represents the universe of matter.

Conclusion

As a spiritual interpretation of a scientific theory, the *raison d'être* of the universe is simply to allow our minds to grow in a safe environment where causality is well defined. We are here to learn the difference between right and wrong so that we can exist without the crutch of causality when the time finally comes to break out of our oceanic shell. The really big secret in the middle is that all the world's problems are easily solvable. But for this to happen, we must do one painfully simple painful thing. We must begin to think for ourselves. As the pirate said to the princess in the timeless movie, *The Princess Bride*, "Life is pain my dear. Anyone who says differently is trying to sell you something."

THE GREAT COSMIC ACCOUNTING BLUNDER

SUMMARY → "The Great Cosmic Accounting Blunder" compares the two physical fixed points in the universe—light speed and Planck's constant—and argues that we have been guilty of double counting up until now and that in fact there is but one fixed point which, as it turns out, is the boundary of the universe.

Life is a daring adventure or nothing at all.
    —Helen Keller

I N THE VERY first episode of the television series *Star Trek—The Next Generation,* Captain Jean-Luc Picard of the starship *Enterprise* finds himself in a medieval court forced to defend humanity against the charge of being a dangerous, savage childlike race—as brought forth by the Q continuum, a race of immortal beings, all named Q, who understand everything except mortal existence. The first episode ends with Q agreeing with Picard that man has indeed shown himself to be peaceful and benevolent, and Picard is allowed to go on his way.

DÉJÀ VU

In the very last episode of the television series, which lasted seven seasons, Picard finds himself involuntarily shifting back and forth through three different time periods—the first episode, the last episode, and twenty-five years after the last episode. He encounters an anomaly at one point in space in each of the three time periods. The analysis reveals the anomaly to be a temporal rupture that is growing larger as it travels backwards in time. At this point, Picard is brought out of a deep sleep by the startling revelation that, paradoxically, the analysis itself has caused the rupture. He proceeds to save the day by taking all three *Enterprises* into the anomaly in order to seal the rupture at the focal point. The series ends with the following dialogue between Picard and Q in the same medieval court where the series began.

Picard: I sincerely hope this is the last time that I find myself here.

Q: You just don't get it, do you, Jean-Luc? The trial never ends. We wanted to see if you had the ability to expand your mind and your horizons. And for one brief moment, you did.

Picard: When I realized the paradox?

Q: Exactly. For that one fraction of a second you were open to options you had never considered. That is the exploration that awaits you. Not mapping stars and studying nebula, but charting the unknown possibilities—of existence.

PASCAL'S SPHERE

Consider for a moment two hypothetical spheres existing in abstract metaphysical space—that is, space where the normal rules of physics do not apply. With the first sphere, the center is everywhere and the boundary is nowhere. With the second sphere, the boundary is everywhere and the center is nowhere. The question is, how are the spheres different? This thought problem leads to the counterintuitive conclusion that the terms *center* and *boundary* are interchangeable in this case. Thus, paradoxically, both spheres describe the very same continuum.

1900—THE QUANTA

Max Planck (1858–1947) first encountered the "ultraviolet catastrophe" thought problem while on his way to understanding why we are able to stand so close to a fire without being overwhelmed by radiation. Classical physics tells us the relationship between the temperature of a fire and the amount of radiation ought to increase linearly. Planck realized that the size restriction on escaping energy units causes a "traffic jam." Therefore, we avoid the catastrophe that would occur if energy were allowed to escape the fire in lesser quantities. Strangely, nature insists that we not be permitted to portion energy in units smaller than Plank's constant. The discovery of this cosmic fixed point set in motion a sequence of cascading paradigms that culminated in the hard-fought realization of quantum theory twenty-five years later.

1905—SPECIAL RELATIVITY

John Wheeler remarked once that, "We have time so that everything does not happen at once." And we have space so that we all do not stand in the same place. The spacetime continuum is our playground. Einstein discovered the equivalence of space and time in 1905 when he realized the invariance of light speed contradicts the additivity of velocity. It turns out that space and time dilate as a function of velocity relative to light speed. Thus, Einstein concluded that all spatial and temporal referencing is relative except for light speed. Then, taking the dilation of space and time to the limit, we see that bodies traveling at light speed exist at the very boundary of spacetime—beyond which lies an unimaginable abyss of nothingness.

1915—General Relativity

Consider for a moment a cat and two closed boxes—one box on Earth and the other accelerating though outer space. The question is, would a cat inside either one of the two boxes be able to tell the difference? The answer is *no*. While Galileo (1564–1642) treated gravity and inertia as mathematically equivalent, it was Einstein who realized they are, in fact, the very same thing. Consider now whether an accelerating cat would feel the effects of inertia if the universe were made empty suddenly. The answer is *yes*. According to general relativity, matter grips spacetime, thus giving it mass and providing a sense of inertia in an empty universe. But since we already know that everything is relative except for light speed, we have reason to believe that matter grips light speed rather than spacetime.

1925—Quantum Theory

While relativity theory speaks to the macrocosmos, quantum theory concerns itself with the nature of matter at the microcosmic level. In characterizing quantum theory, consider again a variation of Schrödinger's classic thought problem involving a quantum cat and two closed boxes. Schrödinger argued absurdly that the cat must be in both boxes until one is opened and the cat's location is determined directly. Heisenberg captured the essence of this quantum indeterminism perfectly in 1927 with his famous uncertainty principle—which states that causality breaks down at Planck's constant. This *acausality* means that we cannot track bodies spatially or temporally beyond the spacetime boundary of Planck's constant. But the movement of electrons across the boundary is what actually determines the nature of matter, suggesting that the movement is also the mechanism by which matter holds its place. This remarkable realization compels us to paradoxically conclude that matter grips not only light speed but Planck's constant as well.

No Boundary? No Dice

Blaise Pascal (1623-62) was a mathematical prodigy, religious philosopher, and one of the greatest thinkers of all time. He founded modern probability theory and contributed to the advancement of differential calculus and projective geometry. Consider again Pascal who described the universe as a sphere in which the center is everywhere and the boundary nowhere. Einstein made a similar claim in saying that the universe is closed but unbounded. Unfortunately, this belief in unboundedness prevented Einstein from conceiving the notion of bodies crossing a boundary and leaving spacetime to travel into nothingness where the normal rules of physics do not apply. His claim that God does not play dice offers further evidence that Einstein could not conceive of an adjacent metaphysical space of nothingness devoid of normal physical rules.

DÉJÀ VU TOO

Consider again the two spheres. With the first sphere, Planck's constant is everywhere and light speed is nowhere. With the second sphere, light speed is everywhere and Planck's constant is nowhere. The question is, how are the spheres different? This thought problem leads to the counterintuitive conclusion that the terms *Planck's constant* and *light speed* are interchangeable in this case. Thus, both spheres describe the very same spacetime continuum. Is this what they mean by thinking outside the box?

2001—THE THEORY OF ONE

The single greatest thought problem occupying the world of physics during the past ninety years involved the attempt to unite the macrocosmos of relativity theory with the microcosmos of quantum theory. As evidence that no prior relationship existed between light speed and Planck's constant, John Wheeler's 1999 book, *A Journey into Gravity and Spacetime,* fails to even mention Planck's constant. The theory of one proves that the universe is bounded, that there is only one photon, and that lone photon is God. Clearly, the theory of one resolves this ninety-year-old thought problem in utterly spectacular fashion. And it is at this point that I wish to stake claim to the greatest scientific discovery of all time—that light speed and Planck's constant are the very same boundary of the spacetime continuum.

REALIZING THE THEORY OF ONE

One of the implications of light speed and Planck's constant being the same boundary of the spacetime continuum is that light speed and Planck's constant are all around us and not at some distant star. Electrons are popping in and out of existence right in front of our noses. It means that reality is an illusion as if we were sitting in a movie theatre watching events projected onto the screen. Imagine we are watching the movie, *The Matrix*. The movie tells us that inside the matrix there is no physical reality and consciousness and our bodies only exist metaphysically. I am saying that here reality is an illusion on two different levels—the movie theatre and the movie itself. This means that the moon does not exist when no one is looking at it. Realizing this represents a massive paradigm shift for all mankind.

CONCLUSION

At one point during Picard's daring adventure, Q takes him back 3.5 billion years to when the first two amino acids were getting together to begin the glorious assault on the abyss that is evolution. The reflection on the road traveled from these two original building blocks brings us face-to-face with the realization that our minds and our horizons will expand only if we are at once fully prepared to boldly go where no man has gone before. This realization in turn brings us face-to-face with the thought problem that asks whether life is a daring adventure or nothing at all.

THE ASPECT EXPERIMENT

SUMMARY → "The Aspect Experiment" argues that the 1881 Michelson-Morley experiment provides empirical validation of special relativity theory and, in a fascinating parallel, that the 1982 Aspect experiment provides empirical validation of my theory of one. This chapter concludes by arguing that physicists are lying to us, because a lie of omission is still a lie.

In 1982, the French physicist Alain Aspect performed an experiment which proved the angles of polarization of twin photons were correlated in such a way that it indicated the photons were instantaneously connected to one another. This is a mind-boggling finding. It means that some of our most cherished and accepted notions about reality are in error. However, what is all the more astounding is that the Aspect experiment—an experiment which could change our understanding of reality as much as the revelations of Copernicus and Darwin—went almost completely unnoticed by the mass media.

—*Michael Talbot*

RECENTLY SENT MATERIAL pertaining to my theory of one to *The Big Bang Theory* television show. They failed to respond. The two roommates on the show are Sheldon and Lenard. Sheldon is a theoretical physicist, and Lenard is an experimental physicist. They help to illustrate that in any scientific endeavor, both theory and experiment are necessary to complete the picture. Socrates said we must answer the question of, "What is X?" before we can say anything about X. My theory of one says X = boundedness. Imagine an episode of the show where Sheldon learns the universe is bounded based on a dead simple application of the Pythagorean Form—thereby shifting his paradigm. My question is, why haven't the physicists figured out this obvious fact that the universe is bounded? I would argue that they have chosen the normal science of string theory over the paradigm-shifting science of the theory of one.

EXPERIMENT AND THEORY

Aristotle, the theorist, believed that heavier bodies fall faster than lighter bodies. Galileo, the experimentalist, proved that all bodies in a vacuum fall at the same speed. Here, theory does not concur with experiment. Sir Arthur Eddington (1882–1944) was a British astronomer who was the senior member of an expedition to Africa to test general relativity theory by observing a total eclipse of the sun. The experiment confirmed the bending of light in accordance with the predictions of general relativity theory. While relativity theory speaks to the macrocosmos, quantum theory concerns itself with the nature of matter at the microcosmic level. Based on quantum theory, consider again the Schrödinger's cat thought problem. He put forth his classic thought experiment involving a quantum-cat and two closed boxes. Schrödinger absurdly argued that the cat must be in both boxes until one is opened and its location is determined directly. Here again, theory concurs with experiment.

The Michelson-Morley Experiment

In 1881, two Americans named Albert Michelson and Edward Morley performed a monumentally important experiment which established that light speed is fixed invariably at 186,284 miles per second. Michelson and Morley were attempting to establish how fast the Earth travels through space by measuring light speed in different directions by using a rotating table. They turned the table in different directions in order to determine how light speed varies. It turns out that it does not vary at all. The theory underlying this experiment was that space is a motionless ocean of ether, and the Earth's motion through this ether should be measureable. As it turns out, there is no such thing as ether, and light speed, being the boundary of the universe, is the much sought after fixed point. Just like waves occur in the medium between ocean and air, so does the boundary between spacetime and nothingness act as the medium for waves of both light and matter. The Michelson-Morley result presented a problem in that, according to Newton, velocities are additive, thus contradicting the invariance of light speed.

Special Relativity Theory

Einstein resolved the Michelson-Morley dilemma in 1905 with his relativity theory by revealing that space and time are interrelated quantities. In paralleling Newton, Einstein theorized that the laws of nature are the same for all uniformly moving bodies. But unlike Newtonian physics, which only concerns itself with mechanical laws, relativity theory also accounts for the behavior of light and other electromagnetic radiation. Consider a train traveling at thirty miles per hour. Newton tells us the photon projected from the headlight would be traveling at light speed plus thirty miles per hour. The Michelson-Morley experiment proved that it is only traveling at light speed, because light speed does not vary, so the photon cannot differ from light speed.

The Aspect Experiment

In 1982, a Frenchman named Alain Aspect performed an astonishing experiment that proved undeniably that all photons are connected instantaneously to one another. The experiment involved lasers and superfast optical switches that measured the time it took for the spin of one photon to be signaled to the spin of another photon. The Aspect experiment revealed immediate signaling between photons. The theory of one explains this instantaneous signaling by recognizing the painfully simple fact that there is only one photon in the universe. The Aspect experiment proves unequivocally that the universal boundary of light speed is all around us, which makes the reality presented to us a façade like a movie being projected in a theatre by the electrons that are popping in and out of the universe without limit. Up until 1982, the scientific community believed that, according to relativity theory, there was no such thing as faster-than-light or instantaneous signaling between particles. Light speed was believed to be the speed limit of the universe. Since photons exist outside spacetime, there is no need for more than one of them because the photon touches every point in the universe. This discovery that there is only one photon means that some of our most strongly held views about reality are factually blunders.

THE THEORY OF ONE

The macrocosmos of special relativity theory is the universal law of spacetime and reveals that spacetime dilates as a function of velocity relative to light speed in accordance with the Pythagorean Form—$h^2 + (v/c)^2 = 1^2$, where h = height, v = velocity, and c = light speed. According to relativity theory, if v = c, then h = 0, thereby indicating a boundary of spacetime. Consider if one had a right-angle triangle with three on one side and four on another, then the question is what is the length of the third side? The answer according to the Pythagorean Form is five (i.e. $3^2 + 4^2 = 5^2$). The Pythagorean Form is used in everything from carpentry to relativistic physics. It is interesting to note that photons travel at light speed—thus implying that photons must exist outside the spacetime continuum. As such, one can imagine a fantastic number of photons zipping around the periphery of the universe at light speed. Rather than having all these photons speeding around the universal boundary, I argue that there is only one of them—which is consistent with the Aspect experiment. The elementary picture of the theory of one is represented by a pearl, whereby the universe proper is metaphorically represented by a grain of sand inside the pearl. The pearl substance itself represents a metaphysical field known as heaven, which houses God, souls, and forms. I argue that God is both the photon (i.e. Light) and a mathematician, as per Sir James Jeans' (1877-1946) argument that God is a mathematician. Jeans was a mathematician and friend to Einstein.

THE ILLUSION OF REALITY

In 1607, Galileo discovered that Jupiter has a fourth moon. This discovery reminded the Church that the Earth orbits the Sun and the Garden of Eden is not the bellybutton of the universe. Similarly, the Aspect experiment and my theory of one remind physicists that reality is an illusion, as described in Michael Talbot's 1991 book, *The Holographic Universe*. He wrote that, "There is compelling evidence to suggest that our world and everything in it—from snowflakes to falling stars and spinning electrons—is only projected images from a level of reality so far beyond our own it is literally outside of spacetime." Mystics, tantrics and idealists have propounded for centuries that reality is an illusion. I have now put forth a scientific basis for this paradigm-shifting belief.

The Unpardonable Sin

The unpardonable sin is an offence so treacherous that even God will not forgive it. Christ defined it as, "Irreverence against the Holy Spirit. It is the refusal to follow the argument when seen." Government agents are anyone who receives funding from the government (e.g. educators, doctors and politicians). Government agents, like academic physicists, have tried unsuccessfully to avoid the unpardonable sin by claiming not to see the simple truth that the universe is bounded, as the result of using the head-in-the-sand method of assessing arguments. Jean-Paul Sartre put an existential spin on the unpardonable sin by stating that, "Those who hide their complete freedom from themselves out of a spirit of seriousness or by means of deterministic excuses, I shall call cowards." Existentialism is the belief that each of us has total freedom and total responsibility.

Déformation Professionnelle

Déformation professionnelle is a French term that means the inclination to see things from the viewpoint of one's own profession rather than from a larger, holistic perspective. Physicists are professionally deformed in that they are unable to comprehend the simple truth of the Aspect experiment, which is that all photons are instantaneously connected to one another because there is only one of them. This fact that there is only one photon means that the physical universe exists inside the photon, meaning innate reality is not the same as that which appears to us.

There is only one photon yet it appears to us that there are a vast number. The Aspect experiment exposes what many religious leaders and philosophers have been telling us for centuries—that reality is an illusion. This means the moon does not exist when no one is looking at it. It is this ultimate truth that physicists are simply unable to comprehend. These conclusions were unclear until I made them clear with this book.

BAD FAITH

To quote Jean Wahl from the foreword to Sartre's 1965 essay, *Bad Faith*,

> Bad faith is the term coined by Kierkegaard's wayward twentieth-century disciple Sartre which is a state of human inauthenticity where one attempts to flee from freedom and responsibility. It is a paradoxical and therefore ultimately schizophrenic attempt at self-deception. To live in good faith means to always strive for authenticity and to continually be aware of the tendency to slip into bad faith.

Saint Augustine said, "The first step forward is to see that attention is fastened firmly on the truth." Physicists are living in bad faith and have a different philosophy of being. For them, the first step forward is to see that attention is fastened firmly to their careers. As Heisenberg tells us, "The history of physics teaches us that the abandonment of earlier concepts is much more difficult that the adaptation of new ones." Physicists are unable to let go of string theory, their outdated meal ticket.

Conclusion

The Michelson-Morley experiment proves that light speed is invariably fixed, which provides empirical validation of special relativity theory. In an amazing parallel, the Aspect experiment proves that all photons are connected instantaneously to one another, which provides empirical validation of my theory of one photon. I have judged physicists on three moral standards—the unpardonable sin, *déformation professionnelle*, and bad faith. They have failed them all. The problem is that physicists cannot wrap their minds around the fact that reality is a façade. In fact, physicists are not physicists at all. They are just people who behave like physicists are supposed to behave. They write all sorts of complicated equations on the board but are unable to simply describe innate reality.

THE THEORY OF ONE

SUMMARY → "The Theory of One" character-
izes relativity theory, quantum theory, and
the theory that unites relativity theory and
quantum theory—the theory of one.

Quantum theory does not hold undisputed
sway, but must share dominion with that other
rebel sibling—relativity theory. And although
these two bodies together have led to the
most penetrating advances in the search for
knowledge—they must remain enemies. Their
fundamental disagreement will not be resolved
until both are subdued by a still more powerful
theory that will sweep away our present painfully won fancies concerning such things as
space, time, matter, radiation and causality. The
nature of this theory may only be surmised—but
it will ultimately come down to the very same
certainty as to whether our civilization as a
whole survives—no more no less.

—*Banesh Hoffmann*

ÉVARISTE GALOIS (1811–1832) was a French mathematician who founded modern group theory and made significant contributions to the theory of algebraic equations—although virtually nothing he did was understood in his lifetime. Group theory is the basic structure of modern algebra, consisting of a set of elements and operators. It plays a central role in both relativity theory and quantum theory. In 1830 Galois was expelled from the academy for scorning the staff and students for their lack of backbone. Galois submitted an anonymous paper on the general solution of equations that is now called "Galois theory," but which, at the time, was described as incomprehensible. In 1831, Galois was arrested for speaking against the king and wearing an illegal uniform, for which he received six months in prison. Upon his release he was seduced by a police-sponsored prostitute and then challenged to a duel over the prostitute by a police agent. Galois spent the night feverishly sketching out as many of his mathematical discoveries as he could, occasionally breaking off to scribble in the margin, "I have not enough time." At dawn, he received a pistol shot in the stomach and was left where he fell. He was buried in a common ditch eight days later at the age of twenty. Galois reminds us how good we have it today in most countries. The government in Canada will not answer any of my arguments, but they have not imprisoned me for making them.

CLASSICAL PHYSICS

One hundred and twenty years ago, it was believed that physics was on the cusp of conclusion. Comprehensive theories were available for characterizing the two known forces, gravity and electromagnetism. Newton's laws described the mechanics of physical bodies, and Maxwell depicted the wave mechanics of light and other electromagnetic forces. Scientists believed the universe was a deterministic mechanism that adhered to causality. Laplace claimed that we should be able to perfectly predict the future if we have detailed knowledge of the present. Then, in 1897, determinism, causality, and classical physics began scrambling when Thomson discovered the electron and the divisibility of the seemly solid atom. He set in motion a series of falling domino-like paradigms that resulted in the revelation of quantum theory in 1925.

RELATIVITY THEORY

In 1881, Michelson and Morley, performed an experiment that established that light speed is fixed at 186,284 miles per second. In 1905, Einstein revealed that space and time are, in fact, the combined concept of spacetime. With his special relativity theory, he revealed that spacetime dilates as a function of velocity relative to light speed in accordance with the Pythagorean Form, which states that if one knows the length of two sides of a right-angle triangle, then one can calculate the length of the third side. The proof by Pythagoras (571–495 BC) is a landmark in our civilization. According to Newtonian physics, velocities are additive, so a baseball thrown forward at seventy miles an hour from atop a train traveling at thirty miles an hour would be traveling at a hundred miles an hour. The reader might wrongly believe that a photon projected from the headlight of the train to be traveling at light speed plus thirty miles per hour. Relativity theory proves mathematically that the photon would only be traveling at light speed. This is because the train begins to dilate microscopically once in motion. Consider an example of relativity theory with an astronaut traveling at 87 percent of light speed. Time would elapse at half his original rate, while he would appear to be half his original height, according to the Pythagorean Form. By taking the dilation of spacetime to the limit, we see that astronauts traveling at light speed would shrink to zero height and, therefore, would exist at the boundary of spacetime. Essentially, what Einstein did with relativity was to encapsulate Newtonian physics into Maxwellian wave mechanics. This enables us to understand how dimensions compress and mass increases as physical bodies accelerate toward light speed. By following this line of thought through to its logical conclusion, Einstein realized that energy and matter are simply different forms of the same substance, as depicted by his famous $E = mc^2$ equation, just as gravity and inertia are different forms of the same thing.

QUANTUM THEORY

While the practical applications of relativity theory are limited to the study of the universe at large, the quantum theory of the atom is the basis for the periodic table and all electronic equipment, such as televisions, computers, and laser disk players. In 1900, Max Planck discovered that energy is transferred in discrete packets or *quanta*, as defined by Planck's constant (i.e. E = hv, where h = 6.626 x 10^{-34} Joule-seconds, E = energy, and v = frequency). Planck realized that the size restriction on escaping energy units causes a quantum traffic jam. In 1911, Rutherford recognized that both the solar system and the atom have nuclei containing about 99.9 percent of the mass while occupying about one-billionth of the spherical space. In 1925, Schrödinger proposed an atomic model based on a simple wave equation—If one imagines dropping a pebble in the ocean, then the ripples become the valance rings of the orbiting electrons. The unexpected surprise was that the waves represent the probability of finding an electron at any given point, with the wave crests representing the highest probabilities. So disgusted was Schrödinger with this probabilistic interpretation of his wave equation that he formulated his classic cat in a box thought problem in 1935 with the intention of demonstrating the absurdity of the probabilistic interpretation once and for all. In fact, Schrödinger's cat thought problem revealed itself to be of paramount importance in understanding the monolithic importance of quantum theory. Schrödinger argued that for the probabilistic interpretation to be true, the cat must be both alive and dead until the box is opened and the observer's conscious-ness determines the fate of the cat. The thought problem proves that the cat is both alive and dead until observed and that consciousness is the deciding factor. Even today, physicists focus exclusively on practical applications and avoid the topic of consciousness altogether. The physicist Stephen Hawking once said, "Every time someone men-tions Schrödinger's cat, I go for my gun." Physicist John Polkinghorne concurred, saying, "Your average quantum mechanic is about as philosophically minded as your average garage mechanic."

THE THEORY OF ONE

According to relativity theory, a body traveling at light speed exists at the boundary of spacetime. Heisenberg captured the essence of quantum indeterminism perfectly in 1927 with his famous uncertainty principle, which states that causality breaks down at the spacetime bound-ary of Planck's constant. Causality is simply an ordering of time. An absence of causality means an absence of time, which, in turn, indicates a boundary of time. In 1915 Einstein revealed that gravity and inertia are the same thing by proving the surface of a sphere and a circle are mathematically equivalent (i.e. a one-to-one mapping). To understand this, imagine a sphere sitting atop a flat surface. By drawing a line from the north pole through every point on the equator and below, a circle is created on the flat surface. Then, by doubling the area of the circle, we can see that the sphere and circle are mathematically equivalent. Pascal described the universe as a sphere in which the center is everywhere and the boundary is nowhere. Einstein also said that the universe is unbounded. Consider a version of Pascal's sphere in which Planck's constant is everywhere and light speed is nowhere—and the inverted sphere in which light speed is everywhere and Planck's constant is nowhere. Consider now a flat surface characterizing the universe of all universes. Our universe occupies no more than a point in the universe of all universes, because space and time have no meaning outside of our universe. From this we can say that every point in our universe is both at the centre and the boundary of the universe of all universes. It follows that Pascal's sphere and the inverted Pascal's sphere are mathematically equivalent. And Planck's constant and light speed are the same centre and boundary of the spacetime continuum.

Conclusion

It is well established that the greatest scientific problem of all time is how to marry relativity theory with quantum theory. Relativity theory is the natural law of spacetime and is based on light speed. It describes spacetime dilation in accordance with the Pythagorean Form. The quantum theory of the atom is the natural law of matter and is based on both Planck's constant and a probabilistic wave equation. I have solved the problem of how to unite relativity theory and quantum theory with my theory of one by recognizing that light speed and Planck's constant are the very same boundary of the spacetime continuum.

THE THEORY OF ONE REVISITED

SUMMARY → "The Theory of One Revisited"
delineates seven arguments in support
of my theory of one. They are: Inside Out,
Outside In, The Pythagorean Form, Causality
Breakdown, The Aspect Experiment,
Simplicity and Beauty, and Unchallenged. It
concludes by arguing that I have made my
case for the theory of one.

Nobody wants to believe the truth is as
simple as it is.

   —*Stephen Hawking*

N 1998, I began consulting to the chief financial officer (CFO) and treasurer of Canadian Pacific Limited (CPL). I conducted mathematical analysis, wrote valuation reports, and built risk management models. I also designed and administered a Delphi questionnaire for the executives to establish fundamental corporate values. (The Delphi questionnaire uses an iterative approach and is designed to achieve convergence on values.) On January 1, 2001, I launched my Philosophymagazine.com website and also produced a mock-up of *The Theory of One* book, which includes *The Great Cosmic Accounting Blunder* essay, from which the first argument of this book is borrowed. At the exact same time, CPL was breaking up into its five subsidiaries. I showed it to the CFO and treasurer but they were working twelve-hour days. I lost it on account of the fact that I had solved the greatest scientific problem of all time with my theory of one (2001), which unites relativity theory (1905) and quantum theory (1925), but nobody cares—not then and not now. Alas, I discovered that I am the only one whose attention is firmly fastened on the truth.

ARGUMENT ONE—INSIDE OUT

Imagine if you can, two hypothetical spheres existing in abstract, metaphysical space—that is, space where the normal rules of physics do not apply. With the first sphere, the center is everywhere and the boundary is nowhere, while with the second sphere, the boundary is everywhere and the center is nowhere. The question is, how are the spheres different? This thought problem leads to the counterintuitive conclusion that the terms *center* and *boundary* are interchangeable in this case. Thus, both spheres paradoxically describe the very same continuum. Relativity theory is based on light speed, and quantum theory is based on Planck's constant. Consider again the two spheres. With the first sphere, Planck's constant is everywhere and light speed is nowhere, while with the second sphere, light speed is everywhere and Planck's constant is nowhere. The question is, how are the spheres different? This thought problem leads to the counterintuitive conclusion that the terms *Planck's constant* and *light speed* are interchangeable in this case— and, paradoxically, both spheres describe the very same spacetime continuum.

ARGUMENT TWO—OUTSIDE IN

Consider a tabletop representing the universe of all universes. It is true that our universe occupies no more than a point in the universe of all universes. As such, particles exist deterministically or with some degree of probability at every point in the universe—including at the boundary. Thus, the Outside In argument coincides with the Inside Out argument. We could say that the Big Bang (i.e. the creation of the universe) is occurring at every moment going back to the universe's origin sixteen billion years ago when a photon split into another photon, which split into an electron and a positron (i.e. particles of matter and antimatter). By definition, photons travel at light speed and thus exist at the boundary of the universe. From outside the universe, a single photon appears as a spherical film containing the universe—like a translucent pearl encapsulating a grain of sand—which is the elemental conceptual picture of the theory of one. The elemental conceptual picture of relativity theory is the Pythagorean Form. The elemental conceptual picture of quantum theory is the Schrödinger's wave equation, which resembles the waves that occur if a pebble were dropped into the ocean.

ARGUMENT THREE—THE PYTHAGOREAN FORM

According to William Barrett's 1958 book, *Irrational Man—A Study in Existentialism*, reason was a Greek invention. While the Egyptians used the Pythagorean Form as an empirical rule-of-thumb in building pyramids, it was Pythagoras who proved it to be a mathematical truth. In fact the Pythagorean Form was the first realization of reason. The macrocosmos of relativity theory is the universal law of spacetime and reveals that spacetime dilates as a function of velocity relative to light speed in accordance with the Pythagorean Form—i.e. $h^2 + (v/c)^2 = 1^2$, where h = height, v = velocity, and c = light speed. According to relativity theory, if v = c, then h = 0, thereby indicating a boundary of spacetime. On the other hand, according to Newtonian physics, if v = c, then h = 1. In other words, h is unaffected by v. In his 1962 book, *Relativity Simply Explained,* Martin Gardner made the same argument that I just made but did not put the rubber to the road in that he failed to conclude that if h =

0, the physical interpretation points to the realization of a spacetime boundary.

ARGUMENT FOUR—CAUSALITY BREAKDOWN

The microcosmos of quantum theory (i.e. the universe inside the atom) is the universal law of matter and is based on Planck's constant. Causality is simply the ordering of spacetime and is the foundation of the decision-making process. First, the baseball is thrown, and then it breaks the window—not the other way around and not occurring simultaneously. Quantum theory states that causality fails at the spacetime boundary of Planck's constant. An absence of causality means an absence of spacetime, thereby indicating a boundary of spacetime. My theory of one argues that light speed and Planck's constant are the same boundary of the universe (i.e. the spacetime continuum). Inside the universal boundary lies spacetime while outside the boundary lies nothingness.

THE MICHELSON-MORLEY EXPERIMENT

The Michelson-Morley experiment established that light speed is fixed at 186,284 miles per second. The experiment presented a problem in that, according to Newtonian physics, velocities are additive, thus contradicting the invariance of light speed. Einstein fixed this dilemma by revealing that space and time are variable, interrelated quantities. In 1909 Hermann Minkowski said, "Space and time separately have vanished into mere shadows—and only a combined notion of the two preserves any reality." Minkowski fundamentally rebooted the nature of space and time into spacetime. In paralleling Newton, Einstein theorized that the laws of nature are the same for all uniformly moving bodies. But unlike Newtonian physics, which only concerns itself with mechanical laws, relativity theory also accounts for the behavior of light and other electromagnetic radiation. The Michelson-Morley experiment presented a paradigm shift in going from the variability of light speed to its invariability.

ARGUMENT FIVE—THE ASPECT EXPERIMENT

In 1982, Aspect performed an astonishing experiment, which proved that all photons in the universe are connected instantaneously to one another. The theory of one explains this by recognizing the fact that there is only one photon in the universe. Previously, the scientific community believed that there was no such thing as faster-than-light or instantaneous signaling between particles. Light speed was and still is believed to be the speed limit of the universe. The Aspect experiment presents a paradigm shift in going from light speed signaling to the appearance of instantaneous signaling. Quantum entanglement is a complicated phenomenon that recognizes the instantaneous transmission of signals between particles. However it does not explain this effect as the theory of one does—which is that there is only one photon. Both Michelson and Morley and Aspect used advanced equipment to conduct their experiments. The

Michelson-Morley experiment provided empirical validation of relativity theory while the Aspect experiment provided empirical validation of the theory of one.

ARGUMENT SIX—SIMPLICITY AND BEAUTY

As discussed earlier, Ockham's principle of economy states that if all things are equal, the simplest theory tends to be the correct one. In accordance with Ockham's razor and Dirac's principle of aesthetic value, the theory of one is simple, beautiful and reasonable, and therefore, true. It is the theory that light speed and Planck's constant are the same boundary of the universe, that there is only one photon, and that lone photon is God. It is my assertion that if an argument that is simple, beautiful and reasonable it is true by definition.

ARGUMENT SEVEN—UNCHALLENGED

I first published the theory of one on Philosophymagazine.com on January 1, 2001. The theory has gone unchallenged ever since. I have sent out my theory of one material to dozens of government agents across North America, including physicists (see the Government Correspondence section on Philosophymagazine.com). Special relativity theory only took four years to be recognized by the scientific community. Quantum theory was recognized almost immediately. To quote Lewis Carroll (1832–1898), "The time has come, the Walrus said, to speak of many things."

In 1909 Hermann Minkowski (1864-1909) presented a geometric interpretation of relativity theory, and the scientific community took notice. Ironically, Minkowski was also Einstein's university professor and described him as a lazy dog that never bothered with mathematics at all—which makes sense given that Einstein sought elemental conceptual pictures first before considering mathematical complexities.

Conclusion

My theory of one is supported by seven significant arguments that have stood the test of time. As Shakespeare said, "This was something of a paradox for which time now gives its proof." We need to base our society on arguments rather than opinions. All I am asking from the government (e.g. educators, doctors and politicians) is to answer my arguments. You can compel the government to respond to my theory of one by nominating me for the Nobel Prize.

QED.

QUOTATIONS

The greatest discoveries in science have been those which force us to rethink the universe and our place in it.

   —*Robert Park*

If I have seen farther than others, it is because I have stood on the shoulders of giants.

   —*Sir Isaac Newton*

If I have seen farther than others, it is because I am surrounded by dwarfs.

   —*Murray Gell-Man*

The violent reaction to the recent development of modern physics can only be understood when one realizes that here the foundations of physics have started moving—and that this motion has caused the feeling that the ground would be cut from science.

—*Werner Heisenberg*

In characterizing Planck's constant—it is as if we are able to drink a pint of beer or no beer at all. Nature strictly prohibits us from drinking any quantity in between.

—*George Gamow*

If the reader wonders why my book does not include a chapter on the philosophical consequences of relativity, it is because I am firmly persuaded that in the ordinary sense of the word *philosophical*—relativity has no consequences. As far as the great traditional topics of philosophy are concerned—God, immortality, free will, good and evil, and so on—relativity has absolutely nothing whatsoever to say.

—*Martin Gardiner*

During the period from 2020 to 2050 we are likely to enter the fourth phase of computing, when intelligent automatons begin to walk the Earth, and populate the internet. Beyond 2050 we are likely to enter the fifth phase of computing, with the beginnings of robots with consciousness and self-awareness.

—*Michio Kaku*

Superstring theory is a miracle, through and through. It will dominate the world of physics for the next fifty years. We are witnessing a revolution in physics as great as the birth of quantum theory.

—*Edward Witten*

Bitter and sweet, warm and cold, as well as all the colors—all of these things exist in opinion and not in reality. What really exist are unchangeable particles or *atoms* and their motion in empty space.

—*Democritus*

All the choir of heaven and furniture of Earth—in a word all those bodies which compose the mighty frame of the world— have not any substance without the mind. So long as they are not perceived by me, or do not exist in my mind or in the mind of any spirit, they have no existence whatsoever.

—*George Berkeley*

The future belongs to those who believe in the beauty of their dreams.

—*Eleanor Roosevelt*

History teaches us that we have never learned anything from history.

—*G.W. Hegel*

Miracles happen, not in opposition to nature, but in opposition to what we know of nature.

   —*Saint Augustine*

In order that we may truly exist, rather than remain in the sphere of the things-seen and things-used, we must quit the inauthentic sphere of existence. Ordinarily, due to our own laziness and the pressure of society, we remain in an everyday world, where we are not really in contact with ourselves. This everyday world is what Heidegger calls the domain of the Everyman. In this domain of Everyman, we are not conscious of our existence. And an awareness of ourselves is only attainable by traversing certain experiences like that of anguish, which puts us in the presence of the background of Nothingness—from which Being erupts.

   —*Jean Wahl*

Modern man has acquired the willpower to carry out his work proficiently without recourse to chanting, drumming or praying. He thoughtfully and skillfully translates his ideas into actions without a hitch—while primitive man was hampered by fears and superstitions at each step along the way. Yet in maintaining his creed, modern man pays the price in a remarkable lack of introspection. He is blind to the fact that, with all his rationality and efficiency, he is possessed by powers beyond his control that keep him restlessly on the run.

—*Carl Jung*

Modern man wants neither God nor Christ— for what he desires is simply the authority of the Church. He wants the physical security of bread, the spiritual security of dogma, and the so-called proof of the existence of miracles. To follow God irrespective of the consequences presents too great a risk. The Church offers up a lighter burden. It serves, selects and explains the truth, forgives sins and bestows upon man the happiness of children. Yet the price is high. Man must surrender his freedom of thought and, indeed, he willingly does so. He no longer serves God as God demands of him, but only as the Church tells him so. God's mysteries and miracles are henceforth monopolized and administered by the Church.

—*William Hubben*

Thou shall have no other god before me.

—*God*

I sometimes wonder what future historians will say of us. It seems to me a single sentence should suffice for modern man—He fornicated and read the papers. After that vigorous definition, the subject will be, if I may say so, exhausted.

—*Albert Camus*

Of whom and of what indeed can I say—I know that! This heart within me I can feel, and I judge that it exists. This world I can touch, and I likewise judge that it exists. There ends all my knowledge, and the rest is construction. For if I try to seize this self of which I feel sure, if I try to define and summarize it, it is nothing but water slipping through my fingers. I can sketch one by one all the aspects that it is able to assume, all those likewise that have been attributed to it, this upbringing, this origin, this ardor of these silences, this nobility or this vileness. But aspects cannot be added up. This very heart which is mine will forever remain undefinable to me. Between the certainty I have of my existence and the content I try to give to that assurance, the gap will never be filled. Forever I shall be a stranger to myself. In psychology as in logic, there are truths but no truth. Socrates' *Know thyself* has as much value as the *Be virtuous* of our confession-als. They reveal a nostalgia at the same time as an ignorance. They are sterile exercises

on great subjects. They are legitimate insofar
as they are approximate.

—*Albert Camus*

There can be no other truth to take off
from this—I think, therefore I exist—i.e. the
Cartesian *cogito*. There we have the absolute
truth of consciousness becoming aware of
itself. Every theory which takes man out of
the moment in which he becomes aware of
himself is, at its very beginning, a theory
which confounds the truth, for outside the
Cartesian *cogito*, all views are only probable,
and a doctrine of probability which is not
bound to a truth dissolves into thin air. In
order to describe the probable, you must
have a firm hold on the true. Therefore,
before there can be any truth whatsoever,
there must be an absolute truth; and this
one is easily arrived at; it is on everyone's doorstep; it is a matter of grasping
it directly.

—*Jean-Paul Sartre*

Men have called me mad, but the question
is not yet settled, whether madness is or
is not the loftiest intelligence—whether
much that is glorious—whether all that is
profound—does not spring from disease of

thought—from moods of minds exalted at the expense of general intellect.

—Edgar Allan Poe

The universe begins when God creates a primordial particle out of nothing. From it matter irradiates spherically in all directions in an inexpressibly great yet limited number of unimaginably yet not infinitely minute atoms.

—Edgar Allan Poe

Philosophy is a battle against the bewitchment of our intelligence by the means of our language.

—Ludwig Wittgenstein

My enemies are those that destroy the world without creating themselves.

—Fredrick Nietzsche

Never accept human reasoning at face value, for it seeks to mask what it fears to confront—the most unpleasant truth of all.

—Fredrick Nietzsche

Something profoundly convulsive and disturbing suddenly becomes both visible and audible with indescribable definiteness and exactness—bringing on the overwhelming feeling that one is utterly out of hand.

Everything occurs without volition—as if by eruption of freedom, independence, power and divinity—thus giving rise to the most immediate, exact and intense form of expression.

—*Fredrick Nietzsche*

You have the look of a man who accepts what he sees because he is expecting to wake up. And you are here because you know something. What you know you can't explain, but you feel it. You've felt it your entire life. That there's something wrong with the world. You don't know what it is, but its there, like a splinter in your mind driving you mad. It is the world that has been pulled over your eyes to blind you from the truth. Like everyone you are a slave. You were born into bondage, born into a prison you cannot smell or taste or touch—a prison for your mind.

—*Morpheus*

In the evolution of scientific thought, one fact has become impressively clear—that there is no mystery of the physical world which does not point to a mystery beyond itself. All highroads of the intellect, all byways of theory and conjecture lead ultimately to an abyss that human ingenuity can never span. For man is enchained by the very condition of his Being, his finiteness and his involvement in nature. The further he extends his horizons, the more vividly he recognizes the fact that, as the physicist Niels Bohr put it, we are both spectators and actors in the great drama of existence. Man is thus his own greatest mystery. He does not understand the vast veiled universe into which he has been cast for the reason that he does not understand himself. He comprehends little of his organic process and even less of his unique capacity to perceive the world about him in his rationality and his dreams. Least of all does he understand his noblest and most mysterious faculty—the ability to transcend himself by perceiving himself in the act of perception. Man's inescapable impasse is that he himself is part of the world that he seeks to explore—his body and proud brain are but mosaics of the same elemental particles that compose the dark, drifting clouds of interstellar space. Man is, in the final analysis, merely an ephemeral confirmation of the primordial spacetime field. Standing midway between macrocosm and microcosm, he finds barriers on every side and can perhaps but marvel, as Saint Paul did nineteen hundred years ago in saying that the world was created by the word of

God so that what is seen is composed of things which do not appear.

—*Lincoln Barnett—concluding paragraph from* The Universe and Dr. Einstein *(1948)*